Laboratory Notebook

Name ______________________________

Email ______________________________

Subject ______________________________

Notebook # ______________________________

Table of Contents

Page	Title	Date

Table of Contents

Page	Title	Date

Title		Date
Name	Partner	

Title		Date
Name	Partner	

Title	Date

Name	Partner

Title		Date
Name	Partner	

Title		Date
Name	Partner	

Title		Date
Name	Partner	

<table>
<tr><td colspan="2">Title</td><td>Date</td></tr>
<tr><td>Name</td><td colspan="2">Partner</td></tr>
</table>

Title		Date
Name	Partner	

Title		Date
Name	Partner	

Title		Date
Name	Partner	

Title		Date
Name	Partner	

Title		Date
Name	Partner	

Title	Date

Name	Partner

Title		Date
Name	Partner	

Title	Date
Name	Partner

Title		Date
Name	Partner	

Title		Date
Name	Partner	

Title		Date
Name	Partner	

Title		Date
Name	Partner	

Title		Date
Name	Partner	

Title		Date
Name	Partner	

Title		Date
Name	Partner	

Title		Date
Name	Partner	

Title		Date
Name	Partner	

Title		Date
Name	Partner	

Title		Date
Name	Partner	

Title		Date
Name	Partner	

Title		Date
Name	Partner	

Title		Date
Name	Partner	

Title		Date
Name	Partner	

Title		Date
Name	Partner	

Title		Date
Name	Partner	

Title		Date
Name	Partner	

Title		Date
Name	Partner	

Title		Date
Name	Partner	

Title		Date
Name	Partner	

Title		Date
Name	Partner	

Title		Date
Name	Partner	

Title		Date
Name	Partner	

Title		Date
Name	Partner	

Title		Date
Name	Partner	

Title		Date
Name	Partner	

Title		Date
Name	Partner	

Title	Date

Name	Partner

Title		Date
Name	Partner	

Title		Date
Name	Partner	

Title		Date
Name	Partner	

Title		Date
Name	Partner	

Title		Date
Name	Partner	

Title		Date
Name	Partner	

Title		Date
Name	Partner	

Title		Date
Name	Partner	

Title		Date
Name	Partner	

Title		Date
Name	Partner	

Title		Date
Name	Partner	

Title		Date
Name	Partner	

Title		Date
Name	Partner	

Title		Date
Name	Partner	

Title		Date
Name	Partner	

Title		Date
Name	Partner	

Title		Date
Name	Partner	

Title		Date
Name	Partner	

Title	Date

Name	Partner

Title		Date
Name	Partner	

Title		Date
Name	Partner	

Title		Date
Name	Partner	

Title	Date
Name	Partner

Title		Date
Name	Partner	

Title		Date
Name	Partner	

Title		Date
Name	Partner	

Title		Date
Name	Partner	

Title		Date
Name	Partner	

Title		Date
Name	Partner	

Title		Date
Name	Partner	

Title		Date
Name	Partner	

Title		Date
Name	Partner	

Title		Date
Name	Partner	

Title	Date
Name	Partner

Title		Date
Name	Partner	

Title		Date
Name	Partner	

Title	Date
Name	Partner

Title		Date
Name	Partner	

Title		Date
Name	Partner	

Title		Date
Name	Partner	

Title		Date
Name	Partner	

Title	Date

Name	Partner

Title	Date

Name	Partner

Title		Date
Name	Partner	

Title		Date
Name	Partner	

Title		Date
Name	Partner	

Title		Date
Name	Partner	

Title	Date
Name	Partner

Title		Date
Name	Partner	

Title		Date
Name	Partner	

Title		Date
Name	Partner	

Title		Date
Name	Partner	

Title	Date

Name	Partner

Title	Date

Name	Partner

Title		Date
Name	Partner	

Title		Date
Name	Partner	

Title		Date
Name	Partner	

Title		Date
Name	Partner	

Title		Date
Name	Partner	

Title		Date
Name	Partner	

Title		Date
Name	Partner	

Title		Date
Name	Partner	

Title	Date

Name	Partner

Title		Date
Name	Partner	

Title	Date
Name	Partner

Title	Date
Name	Partner

Title		Date
Name	Partner	

Title		Date
Name	Partner	

Title		Date
Name	Partner	

Title	Date

Name	Partner

Title		Date
Name	Partner	

Title		Date
Name	Partner	

Title		Date
Name	Partner	

Title		Date
Name	Partner	

Title		Date
Name	Partner	

Title	Date
Name	Partner

Title		Date
Name	Partner	

Title		Date
Name	Partner	

Title		Date
Name	Partner	

Title		Date
Name	Partner	

Title		Date
Name	Partner	

Title		Date
Name	Partner	

Title	Date
Name	Partner

Title		Date
Name	Partner	

Title	Date
Name	Partner

Title		Date
Name	Partner	

Title		Date
Name	Partner	

Title		Date
Name	Partner	

Title		Date
Name	Partner	

Title		Date
Name	Partner	

Title		Date
Name	Partner	

Title	Date
Name	Partner

Title		Date
Name	Partner	

Title		Date
Name	Partner	

Title		Date
Name	Partner	

Title		Date
Name	Partner	

Title		Date
Name	Partner	

Title	Date
Name	Partner

Title	Date

Name	Partner

Title		Date
Name	Partner	

Title		Date
Name	Partner	

Title		Date
Name	Partner	

Title		Date
Name	Partner	

Title	Date

Name	Partner

Title		Date
Name	Partner	

Title	Date

Name	Partner

Made in the USA
Las Vegas, NV
02 March 2024